TEN

why is it important?

Math Concept Books

- ▷ **ZERO IS SOMETHING**
- ▷ **ONE IS UNIQUE**
- ▷ **PRIMES ARE BUILDERS**
- ▷ **SETS**
  What Are They?
- ▷ **POINTS, LINES, AND PLANES**
- ▷ **POLYGONS**
  Points in a Plane
- ▷ **POLYHEDRONS**
  Intersecting Planes
- ▷ **MEASUREMENT**
  How Much? How Many? How Far?
- ▷ **COUNTING SYSTEMS**
  The Familiar and the Unusual
- ▷ **TEN**
  Why Is It Important?
- ▷ **PATTERNS**
  What Are They?
- ▷ **INFINITY**
  What Is It?

A MATH CONCEPT BOOK

TEN

why is it important?

MARNIE LUCE

ILLUSTRATED BY

CHARLES STENSON

Lerner Publications Company
Minneapolis, Minnesota

The Library of Congress cataloged the
original printing of this title as follows:

Luce, Marnie.
Ten: why is it important? Illustrated by Charles Stenson. Minneapolis, Lerner Publications Co. [1969]

48 p. col. illus. 21 cm. (A Math Concept Book)

Discusses the origin of the number system based on ten, explains how it works, and how it is used in the abacus, the metric system, and our money system. Includes study questions with answers.

1. Decimal system—Juvenile literature. [1. Decimal system] I. Stenson, Charles, illus. II. Title.

QA142.L8 511′.1 68-56709
ISBN 0-8225-0580-0 MARC
AC

International Standard Book Number: 0-8225-0580-0
Library of Congress Catalog Card Number: 68-56709

Third Printing 1973

The number system that is customarily used has a base of ten.

Why ten? Why not five or twelve? Why not sixty or twenty?

We will consider one question at a time, for each of these suggested bases has some merit or has been used at some point in the past.

The Romans recognized both five and ten in their system of counting and numeration. A stroke | was used to represent the unit, or one. | | | | | was replaced by V and X was used for ten. C represented 100 and M, 1000. X and V may have been invented as the strokes were crossed out 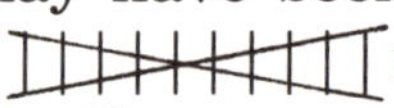 much as is done when counting tally marks. Five tens were indicated by L and five hundreds with a D.

XV

Many people now believe that twelve would be a more convenient number base than ten. Twelve is evenly divisible by 2,3,4, and 6 while ten may only be divided evenly by 2 and 5.

In the base-ten system of numeration there are just ten numerals (0,1,2,3,4,5,6,7,8,9). Therefore, two more numerals would have to be invented to represent ten and eleven in a number system based on twelve.

Base-Ten	Base-Twelve	Base-Ten (cont.)	Base-Twelve (cont.)
0	0	12	10
1	1	13	11
2	2	14	12
3	3	15	13
4	4	16	14
5	5	17	15
6	6	18	16
7	7	19	17
8	8	20	18
9	9	21	19
10	T	22	1T
11	E	23	1E
		24	20

EGGS

Other systems of numeration can be seen throughout the world. An indication of counting by sets of twelve appears in twelve inches to a foot and twelve objects to a dozen. Evidences of a base of sixty are seen in the number of degrees in a circle ($360 = 60 \times 6$), sixty seconds in a minute, and sixty minutes in an hour.

The Mayans of Central America used an efficient number system that had a base of twenty. Probably the word score, meaning twenty, originated with this system.

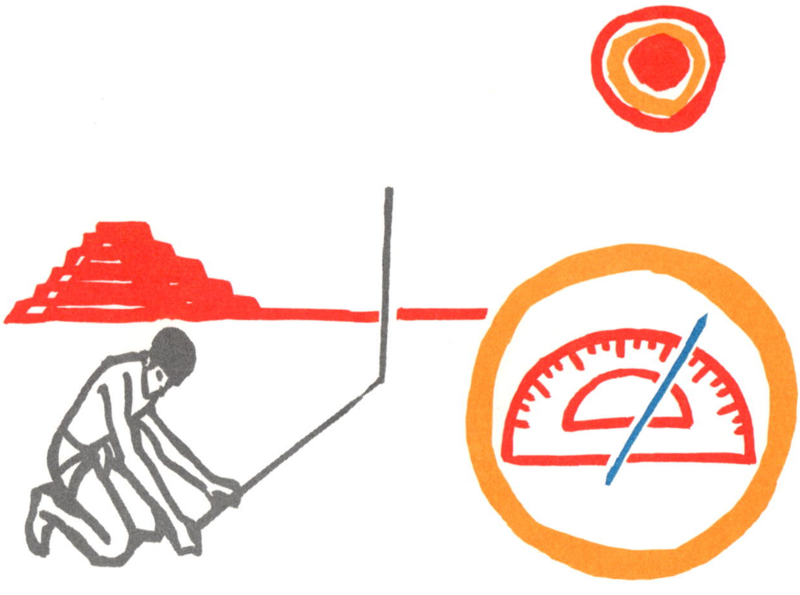

Why ten for our number base?

One answer is immediately apparent—man has ten fingers! If man had only four fingers on each hand a base of eight would probably be used, for fingers were undoubtedly man's first aids in expressing number ideas.

Fingers have been used for other calculations in addition to counting. Men were able to do simple arithmetic with the help of their fingers long before written numerals were devised.

Fingers may be used for a multiplication "machine" if the multiplication facts through 5×5 are known.

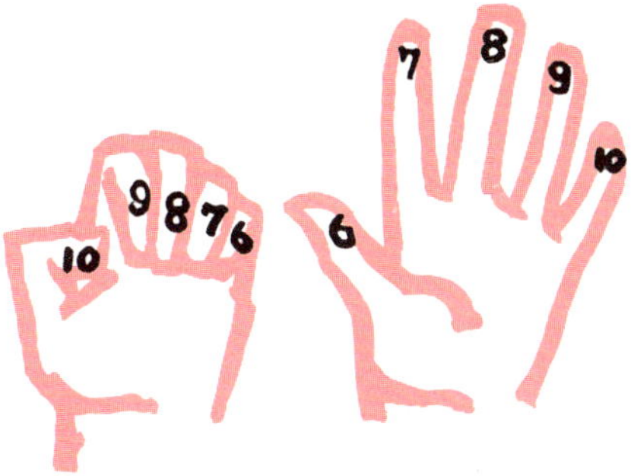

0 × 5 = (5 × 10) + (5 × 0) = 50 + 0 = 50

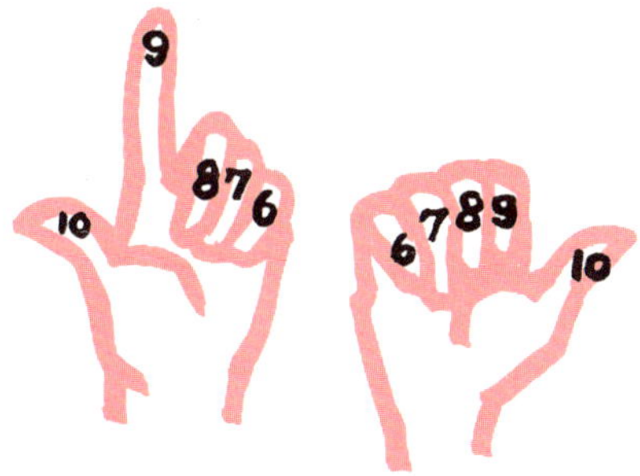

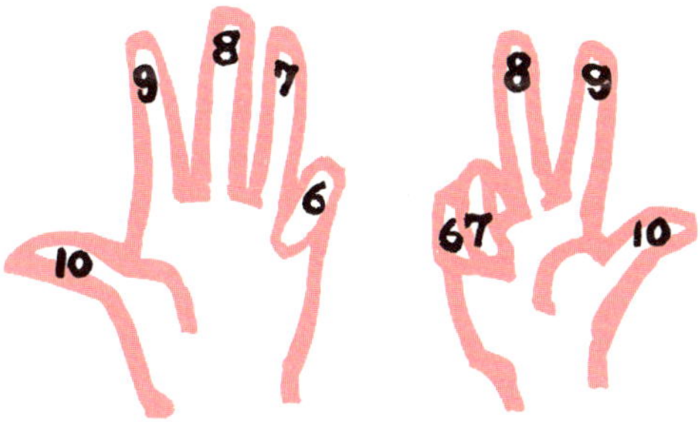

8 × 9 = (7 × 10) + (1 × 2) = 70 + 2 = 72

First, number your fingers as above.

When multiplying six times seven, bend down the "six" finger of one hand and the "six" and "seven" fingers of the other hand. The three fingers which are bent down represent the three tens (30) in the product. Since four fingers of one hand and three on the other hand are upright, multiply 4 × 3 to find the balance (12) of the product.

Therefore, 6 × 7 = (3 × 10) + (4 × 3) = 30 + 12 = 42.

Our present day Hindu-Arabic numerals are the result of many centuries of change and development. The numeral for zero was the last number to be invented.

For centuries computing was done with an abacus or counting-board* and the results were recorded with numerals. Gradually the positional place-value system was devised using 0 as a placeholder. However, a long time passed before numerals and the familiar patterns of computation (known as algorithms) replaced the abacus in problem solving. Even today an expert with the abacus can often solve a problem more quickly than a person working with a pencil and paper.

*The word counter is derived from the counting board.

By using an abacus to solve problems of addition, subtraction, multiplication, and division, we can improve our understanding of our number system.

Since a base-ten system is used, ten beads on a wire may be replaced by one bead on the next wire to the left. Moving from right to left the positions are called units (or ones), tens (10 × 1), hundreds (10 × 10), thousands (10 × 10 × 10), ten thousands (10 × 10 × 10 × 10), hundred thousands (10 × 10 × 10 × 10 × 10), millions (10 × 10 × 10 × 10 × 10 × 10), ten millions (10 × 10 × 10 × 10 × 10 × 10 × 10), etc.

10 ones = 1 ten
10 tens = 1 hundred
10 hundreds = 1 thousand
10 thousands = 1 ten thousand
10 ten thousands = 1 hundred thousand
10 hundred thousands = 1 million
10 millions = 1 ten million
10 ten millions = 1 hundred million
10 hundred millions = 1 billion
etc.

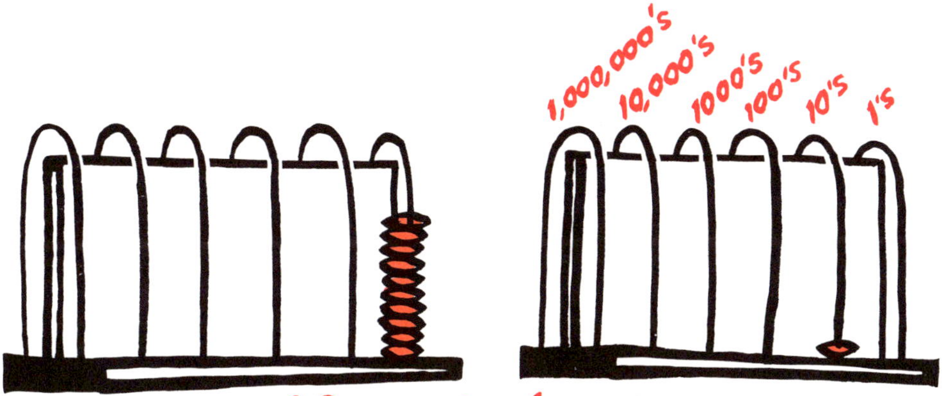

10 ONES = 1 TEN

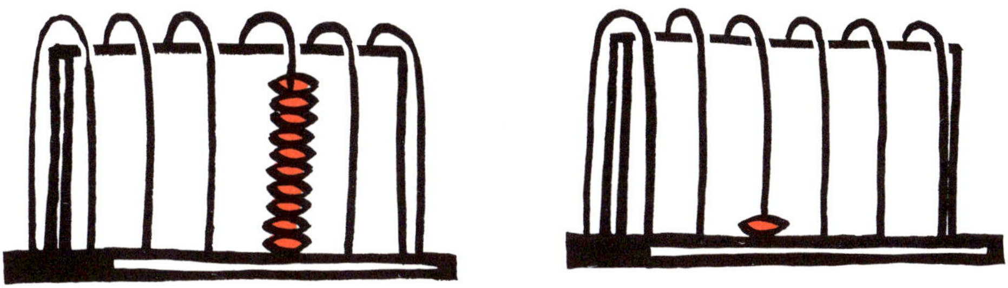

10 HUNDREDS = 1 THOUSAND

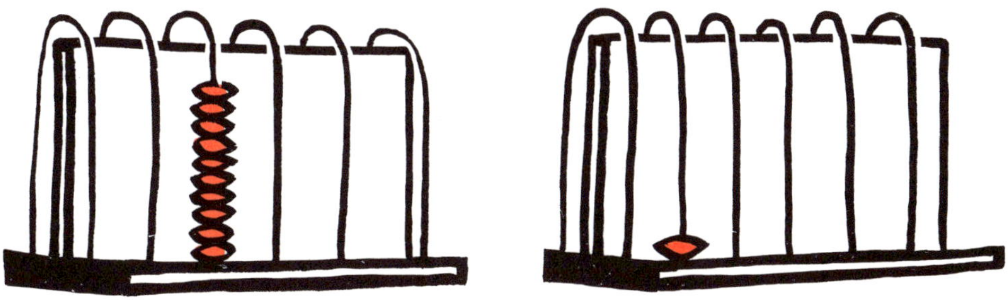

10 THOUSANDS = 1 TEN THOUSAND

To add 524 and 778 the first number to be added (known as an addend) is placed on the abacus by putting 4 beads on the wire at the far right (the ones), 2 beads on the tens wire (moving left one wire), and five beads on the hundreds wire (one place left of the tens wire).

The ones, tens, and hundreds of the addend 778 may be added to the addend on the abacus (524) in any order. However, since it is usually more convenient when writing the algorithm to add the ones first, we will begin there.

After eight beads are added to the 4 on the ones wire, 12 are counted there. Ten of these can be removed and replaced by putting one more on the tens wire. This regrouping is often referred to as carrying the one to the tens.

12 ones = 1 ten and 2 ones

500+20+12

When seven beads are added to those on the tens wire, there are ten. These ten are then replaced by putting another bead on the hundreds wire, which leaves the tens wire empty.

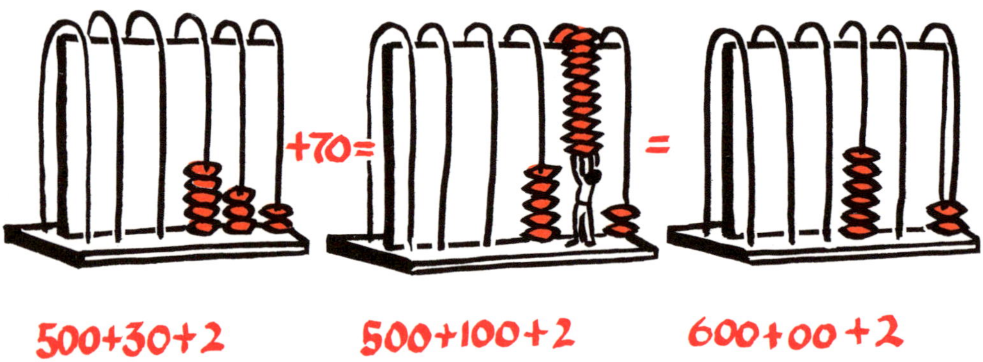

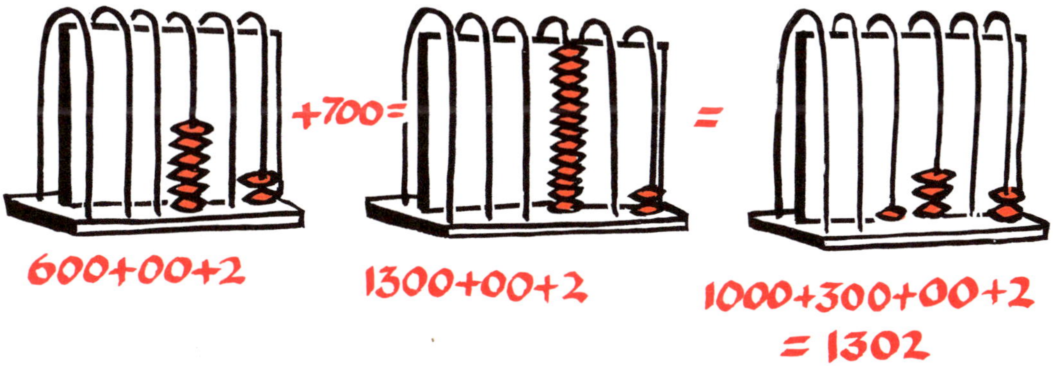

The seven hundreds are added to those already on the hundreds wire which makes thirteen in all. Ten are removed to be replaced by putting one more bead on the thousands wire, which leaves three on the hundreds wire.*

*When computing with the help of numerals, we carry groups of ten to the next greater position instead of moving beads from one wire to another on the abacus.

Let's examine the operation subtraction. The sum 1302 and the addend 778 are given; our job is to find the other addend.

Beads representing the sum 1302 are placed on the abacus.

Obviously eight beads cannot be taken from the two on the one wire. Remember that one ten can be replaced by ten ones and a bead on the hundreds wire can be replaced by ten beads on the tens wire. One bead is removed from the hundreds wire and ten are placed on the tens wires. (This regrouping is sometimes called borrowing.)

One bead is removed from the tens wire which leaves nine. Ten beads are placed on the ones wire which makes twelve in all on the ones wire. Eight ones can now be removed which leaves four ones.

Seven tens (70) can be removed from the tens wire since there are nine beads on that wire. Notice that a numeral may be replaced by a numeral naming a number that is one less than the numeral; this corresponds to the removal of a bead from a wire.

Seven beads cannot be taken from the two beads on the hundreds wire. Therefore one bead is taken from the thousands wire and it is replaced with ten hundreds. Now, seven beads can be taken from the twelve hundreds which leaves five beads.

500+20+4 = = 400+120+4 = 524

When adding and subtracting, sets of ones, tens, hundreds, thousands, etc. may be regrouped; this is sometimes called borrowing and carrying.

Regrouping:

a) How many beads should be placed on the hundreds wire?

4000+0+50+3 = _+_+_+_ = 4053

b) How many beads should be placed on the ones wire?

300+20+5 = ___+___+___ = 325

c) How many beads should be placed on the ones wire?

1000+0+0+0 = ___+___+___ = 1000

Answers page 46

Multiplication may sometimes be regarded as repeated addition. Suppose we want to multiply 27 by 3.

$1 \times 27 = 20 + 7$ $2 \times 27 = 40 + 14$ $3 \times 27 = 60 + 21$

The ones may be regrouped: $21 = 2$ tens $+ 1$

$3 \times 27 = 80 + 1 = 81$

2 is carried to the tens column since 21 ones are equal to 2 tens and 1 one; 21 ones are regrouped as 2 tens and 1 one. 3×2 tens $= 6$ tens. 6 tens $+ 2$ tens $= 8$ tens. Therefore $3 + 27 = 81$.

The number of sets of 9 in 54 (54 ÷ 9) can be found by subtracting 9 from 54 as many times as possible:

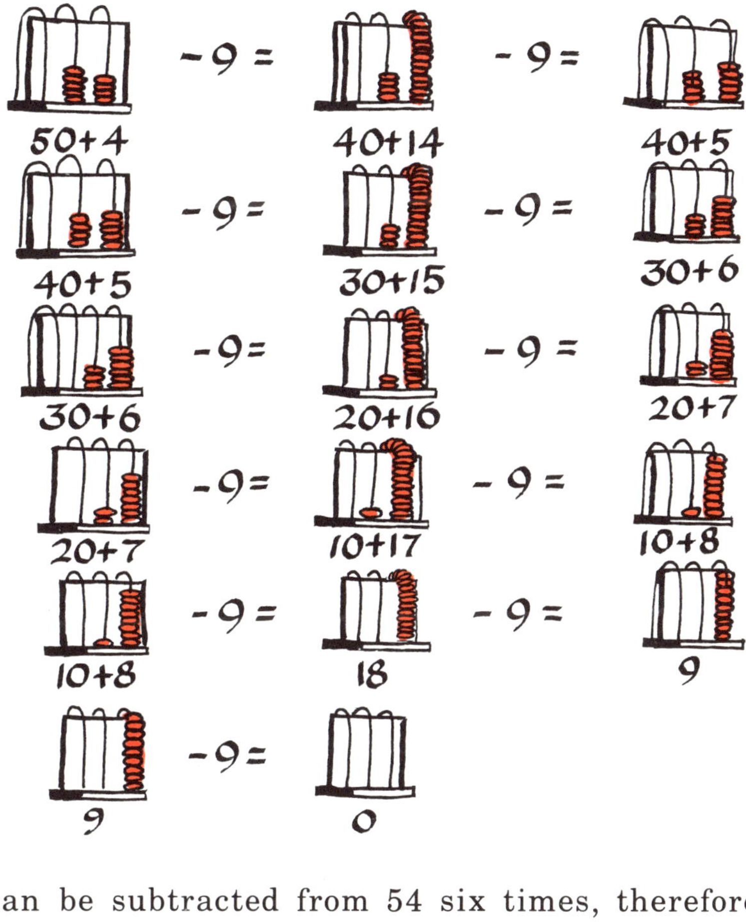

9 can be subtracted from 54 six times, therefore 54 ÷ 9 = 6.

In the base-ten number system the first position on the right is the units, or ones, place. The next place to the left is the tens place: 1 ten = 10 ones. The third place from the right is the hundreds place: 1 hundred = 10 tens (ten sets of ten each).

$10 \times 10 \times 10 \times 10 \times 10 \times 10 \times 10 \times 10$	$10 \times 10 \times 10 \times 10 \times 10 \times 10 \times 10$	$10 \times 10 \times 10 \times 10 \times 10 \times 10$	$10 \times 10 \times 10 \times 10 \times 10$	$10 \times 10 \times 10 \times 10$	$10 \times 10 \times 10$	10×10	10's	Units
1	1	1,	1	1	1,	1	1	1

An exponent may be used to tell how many times the base is used as a factor.

Thus, 100 $= 10 \times 10 = 10^2$ (Read: 10 to the second power or 10 squared.)

1000 $= 10 \times 10 \times 10 = 10^3$ (Read: 10 to the third power or 10 cubed.)

10,000 $= 10 \times 10 \times 10 \times 10 = 10^4$ (Read: 10 to the fourth power.)

Exponents are useful when writing large numbers: for example, 17,000,000,000 may be expressed as 17×10^9. The planet Pluto is about 3,720,000,000, or 3720×10^6 miles from the sun.

Exponents may also simplify the multiplication and division of large numbers:

$$\text{Since } 10{,}000 = 10 \times 10 \times 10 \times 10 = 10^4$$
$$\text{and } 1{,}000{,}000 = 10 \times 10 \times 10 \times 10 \times 10 \times 10 = 10^6$$

$$\text{then } 10^6 \times 10^4 = 10 \times 10 \times 10 \times 10 \times 10 \times 10 \times 10 \times 10 \times 10 \times 10 = 10^{10}$$
$$\text{Since } 1000 = 10 \times 10 \times 10 = 10^3$$
$$\text{and } 100 = 10 \times 10 = 10^2$$
$$\text{then } 10^3 \times 10^2 = 10^5$$

Notice that when numbers are expressed in exponential (EKS po NEN shal) notation using the same base (in this case ten) the product may be found by adding the exponents.

$$10^6 \times 10^4 = 10^{6+4} = 10^{10}$$
$$10^3 \times 10^2 = 10^{3+2} = 10^5$$

Can you discover how exponents simplify the division of large numbers?

$$1000 \div 10 = 10^3 \div 10^1 = \frac{10^3}{10^1} = \frac{10 \times 10 \times 10}{10}$$

$$= \frac{1 \times 10 \times 10}{1} = \frac{10^2}{1}$$

$$= 10^2$$

$$10^3 \div 10^1 = 10^{3-1} = 10^2$$

$$1{,}000{,}000 \div 1000 = 10^6 \div 10^3 = \frac{10^6}{10^3}$$

$$= \frac{10 \times 10 \times 10 \times 10 \times 10 \times 10}{10 \times 10 \times 10}$$

Since $\frac{10}{10} = 1$, we have

$$\frac{1 \times 1 \times 1 \times 10 \times 10 \times 10}{1 \times 1 \times 1} = \frac{10^3}{1} = 10^3$$

$$10^6 \div 10^3 = 10^{6-3} = 10^3$$

To divide two numbers which are expressed in exponential notation using the same number base, the exponents are subtracted.

Complete the following statements:

a) $10^3 \times 10^4 =$ ____ $\times$ ____ $\times$ ____ $\times$ ____ $\times$ ____ $\times$ ____ $\times$ ____ $= 10^{\_}$

b) $10^3 \times 10^4 = 10^{\_+\_} = 10^{\_}$

c) $10^6 \times 10^3 = 10^{\_+\_} = 10^{\_}$

d) $10^{12} \times 10^9 = 10^{\_+\_} = 10^{\_}$

e) $10^3 \times 10^5 = 10^{\_}$

f) $10^9 \div 10^6 = 10^{\_-\_} = 10^{\_}$

g) $10^4 \div 10^1 = 10^{\_-\_} = 10^{\_}$

Answers page 46

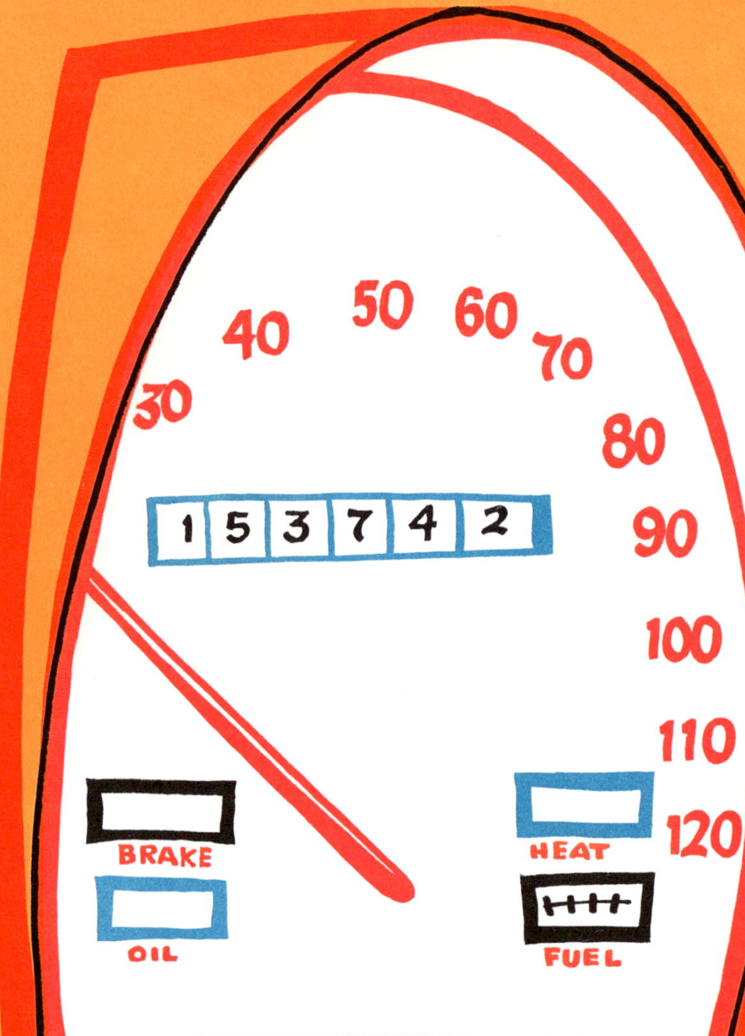
30
40
50
60
70
80
90
100
110
120
153742
BRAKE
HEAT
OIL
FUEL

Have you ever noticed the odometer when you were riding in a car?

In which of the small windows do the numerals change most frequently? The one farthest right?

Often the numerals showing in this place are of a different color than the numerals showing in the other openings. This part of the odometer counts the tenths of a mile as the car is moving along.

How many tenths of a mile are equal to one mile?

Of course, ten-tenths equals one unit whether the unit is a mile, a birthday cake, a candy bar, or a day at the beach!

The odometer above shows that the car has traveled 14,507.6 miles. (Read: Fourteen thousand five hundred seven and six-tenths miles.)

The tenths place will next show 7, followed by 8, 9, and then 0. When 0 shows in the tenths place the 7 in the ones place will change to 8.

a) The numerals in the tenths place will go around to 0 how many more times before 0 shows in the ones place?

b) When 0 appears in the ones place what numeral appears in the tens place?

c) How many times must the tenths numerals go around before the numeral in the ones place is 0 again?

d) When the numeral in the ones place becomes 0 again the 1 in the tens place changes to ____.

e) When the numeral in the tens place changes from 9 to 0 the numeral in the hundreds place will change to ____.

Answers page 47

If the numeral 5 stands alone or in the place farthest to the right (7815) it represents five ones (5×1).

If the numeral 5 is in the second place from the right (50) it is ten times as great as if it were in the ones place; $50 = 10 \times 5$ since ten ones are contained in each of the 5 tens.

From this it is clear that an easy way to multiply a number by ten is to annex a 0 to the numeral: $10 \times 5 = 50$.

a) What is an easy way to multiply by 100?

b) by 1000?

c) by 10,000?

d) by 1,000,000?

Answers page 47

If a small child is offered his choice of a nickel or dime he might take the nickel since it is larger. However, it isn't long before he realizes that although the nickel is larger in size the dime has the greater value.

The monetary system of the United States is easily understood since the basic unit is one dollar which is equal to ten dimes, each dime being equal to ten pennies (cents). The structure is the same as our base-ten number system. Thus, $\$15.32 = (2 \times 1¢) + (3 \times 10¢) + (5 \times \$1) + (1 \times \$10) = 2¢ + 30¢ + \$5 + \$10$. Since 32¢ is 32/100 of a dollar we may write the total as $15.32.

POPCORN

Suppose the man selling popcorn has used most of his change and has only several bills and a few pennies left. If he goes to his friend the balloon vender with a $5 Federal Reserve note (which is usually called a $5 bill), his friend might give him a $1 bill, 3 half-dollars, 2 quarters, 15 dimes, 5 pennies, and the balance in nickels. How many nickels does the balloon vender give the popcorn man?

Answer page 47

Joe and Peter were operating a stand at which they sold lemonade for the following prices: small glass, 3¢; large glass, 7¢; giant-size glass, 10¢. Between them they had a half-dollar, a quarter, and a dime. The neighborhood storekeeper agreed to give them change in nickels, dimes, and pennies.

a) If the storekeeper gave them 2 dimes, 3 nickels, and the balance in pennies how many pennies did he give them?

b) If the storekeeper gave the boys five nickels and the balance in pennies how many pennies did they receive?

c) When the boys had sold all of their lemonade they had 275 pennies, 3 nickels, and a dime. They asked the storekeeper to exchange their coins for the fewest pieces of money that they could divide evenly between them. What pieces of money did he give them in exchange for their 275 pennies, 3 nickels, and 1 dime?

Answers page 48

10¢
7¢
3¢

NO SALE

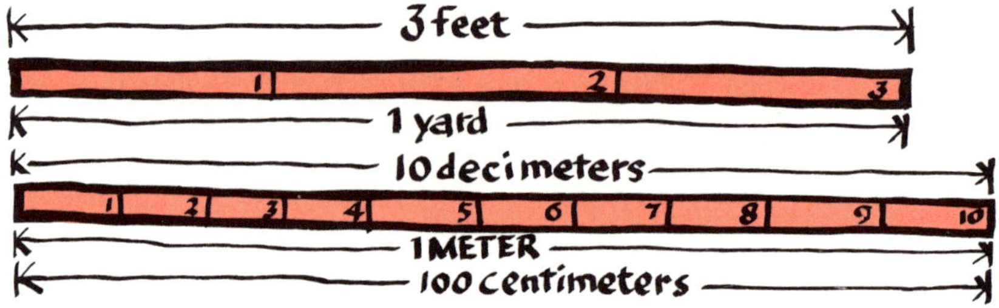

The United States is one of the few countries in the world that does not use a system of measurement based on ten. Scientists in all countries and many industries and governmental agencies of the United States use the metric system of measurement, which is based on ten, because it is more convenient.

To change from one unit to another in the metric system it is only necessary to multiply or divide by a multiple of ten. The various units are related just as the place-value positions of our base-ten number system.

In contrast, if it is necessary to change units of measure in the system used in the United States an irregular series of equivalents must be recalled and used.

Example: To change 3 gallons and 2 pints to pints it is necessary to remember that

4 quarts = 1 gallon
and 2 pints = 1 quart
4×2 = 8 pints in 1 gallon
3×8 = 24 pints in 3 gallons
$24 + 2$ = 26 pints
= 3 gallons and 2 pints

To change 3 kiloliters and 2 liters to liters it is only necessary to remember that the prefix kilo is 1000 × the basic unit (in this case liter). Therefore, 3 kiloliters and 2 liters = (3 × 1000) + 2 = 3002 liters.

Since our number system is based on ten it is most convenient to group and regroup numbers in tens and multiples of ten.

There appears to be but one reason why a base-ten numeration system is used—man has ten fingers.

Answers and Solutions

Pages 26-27:

a) 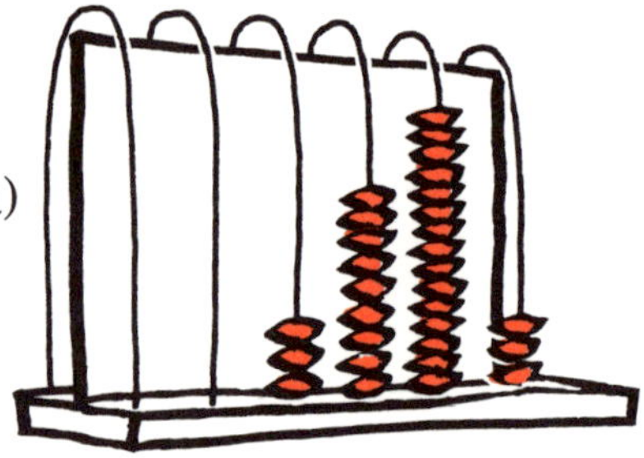

$3000 + 900 + 150 + 3 = 4053$

b) 

$300 + 10 + 15 = 325$

c) 

$900 + 90 + 10 = 1000$

Page 33:

a) $10^3 \times 10^4 = 10 \times 10 \times 10 \times 10 \times 10 \times 10 \times 10 = 10^7$

b) $10^3 \times 10^4 = 10^{3+4} = 10^7$

c) $10^6 \times 10^3 = 10^{6+3} = 10^9$

d) $10^{12} \times 10^9 = 10^{12+9} = 10^{21}$

e) $10^3 \times 10^5 = 10^8$

f) $10^9 \div 10^6 = 10^{9-6} = 10^3$

g) $10^4 \div 10^1 = 10^{4-1} = 10^3$

Page 37:

a) The numerals in the tenths place will go around (1,2,3,4,5,6,7,8,9,0) two times before 0 appears in the ones place.

b) A *1* appears in the tens place.

c) Ten times

d) 2

e) 6

Page 38:

a) Annex two zeros: $100 \times 5 = 500$. (It is not correct to say add two zeros since that would be $5 + 0 + 0 = 5$.)

b) Annex three zeros: $1000 \times 5 = 5000$

c) Annex four zeros: $10{,}000 \times 5 = 50{,}000$

d) Annex six zeros: $1{,}000{,}000 \times 5 = 5{,}000{,}000$

Page 41:

One dollar, 3 half-dollars, 2 quarters, 15 dimes and 5 pennies = (1 × \$1) + (3 × 50¢) + (2 × 25¢) + (15 × 10¢) + (5 × 1¢) = 100¢ + 150¢ + 50¢ + 150¢ + 5¢ = 455¢ or \$4.55. \$5.00 − \$4.55 = 0.45 (500¢ − 455¢ = 45¢) Therefore he gives the popcorn vender 9 nickels since 9 × 5¢ = 45¢.

Page 42:

a) 50¢ + 25¢ + 10¢ = 85¢ (the boys' money)
2 dimes = 20¢; 3 nickels = 15¢; 20¢ + 15¢ = 35¢
85¢ − 35¢ = 50¢ He gave them 50 pennies.

b) 5 nickels = 25¢ 85¢ − 25¢ = 60¢
They received 60 pennies.

c) 275 pennies = 275¢ 3 nickels = 15¢ 1 dime = 10¢
275¢ + 15¢ + 10¢ = 300¢ The storekeeper gave them 2 one-dollar bills and 2 half-dollars.